D0741895

Understanding the Elements of the Periodic Table™

KRYPTON

Janey Levy

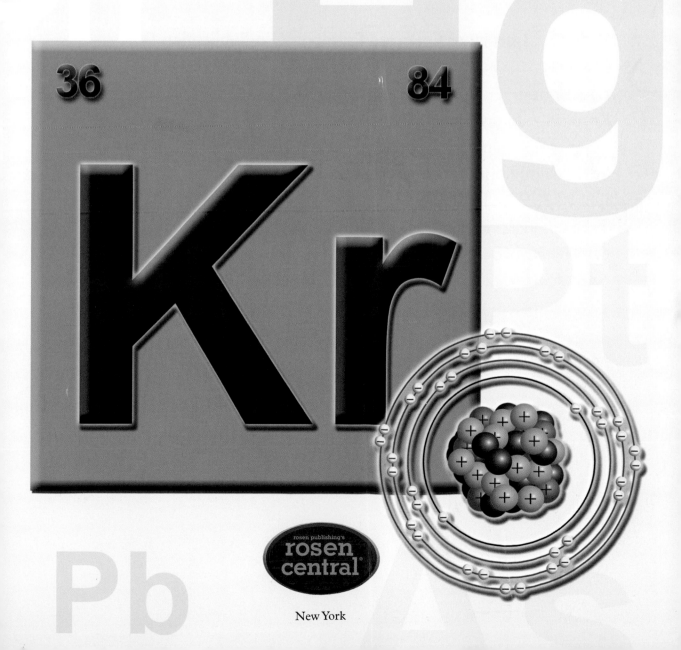

36

84

Kr

rosen publishing's
rosen
central®

New York

Published in 2009 by The Rosen Publishing Group, Inc.
29 East 21st Street, New York, NY 10010
www.rosenpublishing.com

First Edition

Library of Congress Cataloging-in-Publication Data

Levy, Janey.
Krypton / Janey Levy.
     p. cm.—(Understanding the elements of the periodic table)
Includes bibliographical references and index.
ISBN-13: 978-1-4042-1778-2 (lib. bdg.)
1. Krypton. 2. Periodic law—Tables. I. Title.
QD181.K6L48 2008
546'.754—dc22

                                                          2007040005

Manufactured in Malaysia

On the cover: Krypton's square on the periodic table of elements. Inset: The atomic structure of krypton.

Contents

Introduction

The word "krypton" may make you think of Superman. Krypton was the name of his home planet, and kryptonite was the mineral that could drain his superpowers. Given its association with a fictional superhero, it may surprise you to learn there really is a krypton (Kr). It is not a planet, though. It is a chemical element, a gas that belongs to a group of gases called the noble or inert gases.

The discovery of krypton resulted from efforts to answer a scientific puzzle: why was nitrogen (N) obtained from air heavier than nitrogen obtained from chemical compounds? Any scientist will tell you it is common for research into one topic to give rise to questions that lead to new topics. Seeking answers to those new questions is one of the important ways scientific knowledge grows. The nitrogen puzzle was worth solving because of nitrogen's importance. Nitrogen is a colorless, odorless, and tasteless gas that makes up about 78 percent of Earth's atmosphere. It is necessary for life as we know it—all organisms must have nitrogen to live.

Scottish chemist Sir William Ramsay, assisted by British chemist Morris Travers, began a series of experiments in an effort to solve the puzzle about nitrogen. They succeeded, and in the process, discovered krypton.

Their discovery advanced scientific knowledge and gave us more information about the atmosphere that surrounds us. However, it didn't immediately lead to any practical applications for krypton. Scientists

Theodore Gray and Max Whitby build museum displays of chemical elements. They created this version of krypton's square on the periodic table. Handmade discharge tubes form krypton's chemical symbol.

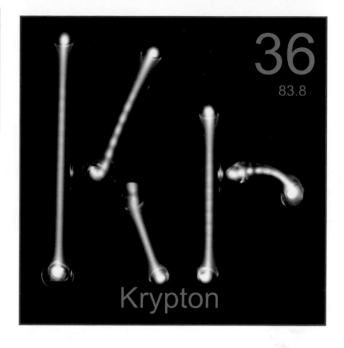

believed that krypton and the other noble gases were inert, or nonreactive. In an article in *The Chemical Educator*, Professor George Kauffman wrote that chemists used to have a joke about the noble gases: if someone wrote a book titled *The Chemistry of the Inert Gases*, it would be filled with blank pages. It was not until more than sixty years after krypton's discovery that scientists learned they were wrong and the noble gases could do interesting things.

In the last half century, scientists have discovered many possible uses for the noble gases. The high cost of obtaining krypton has limited its uses. Still, scientists and ordinary people use krypton in many ways that might surprise you. In fact, it might be helping to provide the light you are using to read this book!

Chapter One
Krypton: The Hidden Gas

Krypton's chemical symbol is Kr. Its name comes from the Greek word *kryptos*, which means "hidden." Sir William Ramsay and Morris Travers called it that because it was so hard to isolate it from the air. They knew some unknown element was present, but it took many attempts before they were able to separate it from other elements in the air. There is nothing special about the "hidden" gas's appearance. It is colorless, odorless, and tasteless. Krypton looks just like air. Krypton is harmless. The only way it could hurt you is if you were to breathe pure krypton and nothing else. Even then, it is not really the krypton that hurts you—it is the lack of oxygen.

Scientists have discovered krypton is useful in many ways. It is used in lighting, lasers, and medicine. It is used to detect leaks in sealed containers and measure X-rays. Scientists have even used it to determine the age of water locked away in spaces deep underground!

The History of Krypton

Krypton was discovered in 1898. However, the story of its discovery actually began more than a century earlier.

In 1785, British chemist Henry Cavendish was studying the formation of nitrous oxide (N_2O) by passing a spark through a mixture of oxygen

Sir William Ramsay is shown here in his University College laboratory around 1900. He was knighted for his work in 1902 and received the Nobel Prize in Chemistry two years later.

(O_2) and air, which is mostly nitrogen. He noticed that each time he did this, there was always a little chemically nonreactive gas left over. Unfortunately, Cavendish did not do what later scientists might have done. He made no effort to study this nonreactive gas. Neither he nor anyone else realized it, but this was the first evidence of the noble gases.

More than a century later, British physicist Lord Rayleigh (John William Strutt) discovered that nitrogen he extracted from ammonia (NH_3) was less dense than nitrogen he isolated from air. This surprised him. Rayleigh theorized that the nitrogen isolated from ammonia was combined with a lighter gas, and this was what made it less dense than nitrogen isolated from air.

Sir William Ramsay thought the reverse was true. He thought the nitrogen obtained from air was combined with a heavier gas. Ramsay suggested this to Rayleigh, and both men began experiments to isolate this heavier gas. They worked separately but exchanged letters almost daily. Finally, in 1894, they both succeeded in isolating the heavier gas. Since it was nonreactive, they named it argon (Ar), from the Greek word *argos*, which means "the lazy one."

In 1895, Ramsay was trying to find sources of argon in minerals when he discovered helium (He). Helium had previously been known only as

Sir William Ramsay used this spectroscope in his work on the noble gases. A spectroscope separates the light given off by an element into its different colors so that it can be studied.

an element in the sun and had been believed to be a metal. Ramsay's discovery demonstrated that helium was actually a nonreactive gas.

Ramsay believed argon and helium belonged to a previously unknown group of elements. He also believed there were other elements in the group and set out to find them. In 1898, Ramsay and his assistant, Morris Travers, discovered three new nonreactive gases. One was krypton. Ramsay named the second new gas neon (Ne), from the Greek word *neos*, which means "new." He named the third new gas xenon (Xe), from the Greek word *xenos*, which means "strange." For his work, Ramsay received the Nobel Prize in Chemistry in 1904.

William Ramsay, University Professor

William Ramsay was a professor at University College in London, England, from 1887 to 1913. He was popular with both students and other professors. His students called him "the Chief." He encouraged his students to carry out original research. He also urged them to learn glassblowing, a skill he possessed. That way, they could make laboratory equipment whenever it was needed. They could also craft it to the exact specifications required.

Putting the Elements in Order

Ramsay might not have continued his search for more noble gases if it had not been for where argon and helium fit on the periodic table of elements. So, what exactly is the periodic table of elements?

The periodic table is an organization of the elements. Scientists realized the need for a system of organization as more and more elements were discovered. Several attempts to organize the elements were made before a widely accepted system appeared. German chemist Johann Dobereiner made the first attempt to organize the elements in 1817. French geologist A. E. Beguyer de Chancourtois created the first periodic table in 1862. His table looked quite different from the one we use today. The periodic table we use today is based on one created by Russian chemist Dmitry Mendeleyev in 1869.

Mendeleyev was a university professor who could not find a textbook he liked for his classes and decided to write his own. While writing his textbook, he noticed patterns among groups of elements. This observation

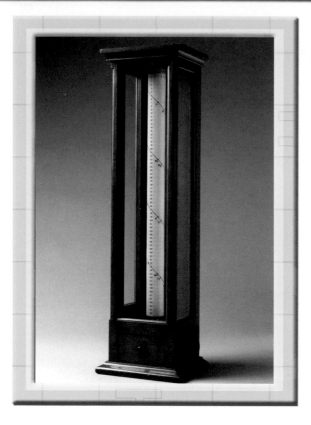

This is the periodic table created by A. E. Beguyer de Chancourtois in 1862. He drew the elements in a continuous line that wrapped around a tube divided into sixteen parts.

led him to discover the periodic law. According to this law, if elements are arranged in order of increasing atomic weight, elements with similar properties appear at regular intervals, or periods. In the periodic table, the elements are arranged in rows of increasing atomic weight, and whenever an element similar to the one at the left appears, a new row is started.

Julius Lothar Meyer

German chemist Julius Lothar Meyer discovered the periodic law about the same time Mendeleyev did. But Mendeleyev announced his discovery first. He also had such confidence in the periodic law that he predicted scientists would discover elements to fill the gaps in the periodic table. Meyer was never brave enough to do that. Since Mendeleyev announced his discovery first and boldly made predictions based on it, he received much more attention than Meyer did.

Following the periodic law, Mendeleyev built up a systematic table of all the elements then known. He arranged the elements in horizontal rows, called periods, and vertical columns, called groups. Elements in a group had similar properties. When an element of the correct properties was not known, Mendeleyev left a gap in the table. Mendeleyev predicted elements would be discovered that would fill those gaps. Based on where the gaps were in the table, Mendeleyev was even able to correctly predict the properties of many of those still-unknown elements.

Mendeleyev's table contained far fewer elements than the modern periodic table. The noble gases, including krypton, had not yet been discovered, and there was no place for them on that early table. However, without the periodic law on which the table is based, the discovery of argon and helium might not have led Ramsay to search for more noble gases.

Chapter Two
Krypton, Atomic Structure, and the Periodic Table

Everything in the world is made up of atoms. There are more than a hundred different kinds of atoms. In an element, all the atoms are alike. Atoms are extremely tiny bits of matter. Krypton atoms are so small that it takes about 400,000 of them placed side by side to equal the width of a single human hair! As tiny as atoms are, they are composed of even smaller bits of matter called subatomic particles.

Protons, Neutrons, and Electrons

Atoms are sometimes described as tiny solar systems. They have a heavy center object with light objects orbiting it. The heavy center object is the nucleus. It is made up of protons and neutrons. The positively charged protons give the nucleus a positive charge because the neutrons have no electrical charge. Together, the protons and neutrons give the atom nearly all its mass. The number of protons determines which element an atom is.

The light objects orbiting the nucleus are electrons. Electrons have a negative charge, so they are attracted to the positively charged nucleus. Usually the number of electrons equals the number of protons, so the atom as a whole has no electrical charge.

The electrons are arranged in layers, or shells, around the nucleus. The number of shells in an atom depends on the number of electrons it

Krypton's thirty-six electrons circle the immense nucleus in four shells. Each of the first three shells is full. The fourth shell, which can hold thirty-two electrons, has eight.

has. Each shell can only hold a certain number of electrons. The first shell can hold two electrons, the second can hold eight, the third can hold eighteen, and the fourth can hold thirty-two. In theory, the three remaining shells can hold many more electrons. In the real world, these outer shells are never filled. All krypton atoms contain thirty-six protons. Most krypton atoms contain forty-eight neutrons, but some have forty-two, forty-four, forty-six, forty-seven, or fifty neutrons. A krypton atom also contains thirty-six electrons, which are arranged in four shells.

What Happens If You Change the Atomic Structure?

Different changes to an atom's structure affect it in different ways. For example, an atom becomes an ion by gaining or losing an electron. Losing an electron produces an ion with a positive charge. Gaining an electron produces an ion with a negative charge. Becoming an ion does not change what element an atom is—that is determined by how many protons it contains. Changing the number of electrons just changes the atom's electrical charge.

Changing an atom's number of protons changes everything. Having thirty-six protons is what makes a krypton atom krypton. If you added a proton to krypton's nucleus, it would be the metal rubidium (Rb). Unlike nonreactive krypton, rubidium is highly reactive. It can suddenly burst into

Electron Energy

Electrons in different shells have different amounts of energy. The ones in the shell closest to the nucleus have the least energy. Electrons with more energy are in the outer shells. The greater energy gives the outer electrons more resistance to the attraction exerted by the positively charged nucleus. As a result, they are more easily removed from the atom.

flames, and it reacts violently with water. If you removed a proton from krypton's nucleus, it would be bromine (Br). Bromine is a highly reactive liquid that readily turns into a gas. It is poisonous, and the liquid can cause severe skin burns. Because the number of protons determines an atom's identity, that number has a special name. It is called the atomic number. In the periodic table on pages 40 and 41 of this book, each element's atomic number appears on the upper left of the element's symbol. Because krypton has thirty-six protons, its atomic number is 36.

Unlike changing the number of protons in an atom's nucleus, changing the number of neutrons does not change what element an atom is. It changes the isotope of the element. Most elements have several isotopes. Each isotope is usually referred to by the element's name plus its mass number, which is the number of protons plus the number of neutrons. Krypton has six naturally occurring isotopes. The most common is krypton-84, which has thirty-six protons and forty-eight neutrons (36 + 48 = 84). Krypton also occurs in about twenty radioactive isotopes.

Because isotopes have different numbers of neutrons, each isotope has a different atomic weight (expressed in atomic mass units, or amu). To determine the official atomic weight of an element, scientists average

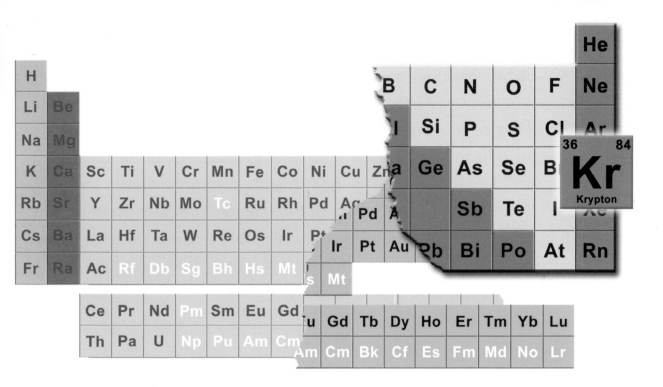

Krypton is the fourth element from the top in the group of noble gases. Above it are helium (He), neon (Ne), and argon (Ar). Below it are xenon (Xe) and radon (Rn).

the atomic weights of all the element's naturally occurring isotopes, taking into account the proportions in which the isotopes occur. On the periodic table in this book, each element's atomic weight appears on the upper right of the element's symbol. Krypton's atomic weight is 83.80, which in our periodic table has been rounded to 84.

Learning About Krypton from the Periodic Table

We can learn more about krypton just from its place on the periodic table. Unlike Mendeleyev's periodic table, which arranged elements according to atomic weight, today's periodic table organizes elements in order of increasing atomic number. This arrangement makes even clearer the patterns of the periodic law. You can use these patterns to help you determine an element's properties.

One of the easiest characteristics you can determine about an element from the periodic table is whether it is a metal, nonmetal, or metalloid (an element with some properties of both metals and nonmetals). You can establish this distinction by looking at the element's relation to the staircase line. The colored blocks on the periodic table create a series of steps, or stairs, on the right side. This is the staircase line. Elements to the left of this line are metals, and elements to the right are nonmetals. Elements

The distinctive properties of metals are clearly visible in this photograph. Copper is in the center, surrounded by aluminum *(left)*, nickel-chrome ore, nickel, titanium, iron-nickel ore, niobium, and chromium.

touching the staircase line are metalloids. Krypton, being a nonmetallic gas, is to the right of the line.

Being a nonmetal means krypton does not share the distinctive properties of metals. Most metals are solids that can be polished to make them shine. They conduct electricity. They are malleable, which means they can be hammered into shapes without breaking. They are ductile, which means they can be stretched into wires.

Look at the periodic table again. You will see it is organized into rows and columns. The rows are called periods and are numbered 1 through 7. Elements in each period have the same number of electron shells. Elements in period 1 have one shell, elements in period 2 have two shells, and so on.

The columns in the periodic table are called groups. Like the periods, the groups are numbered. There are eighteen groups and two systems for numbering them. The traditional system uses Roman numerals combined with letters of the alphabet. The newer system uses Arabic numerals. The elements in each group usually have the same number of electrons in their outer shell. The electrons in the outer shell are called valence electrons, and they have a lot to do with an element's chemical behavior. The elements in each group behave similarly because (usually) they have the same number of valence electrons.

Krypton is found in the group numbered 0 in the traditional numbering system or 18 in the newer system. Other elements in this group are helium, neon, argon, xenon, and radon (Rn). These elements are known as the noble gases. All except helium have eight electrons in their outer shell. Helium has only two electrons, so its one and only shell has two electrons.

Krypton and the Noble Gases

The noble gases' number of valence electrons makes them stable, or unlikely to react with other atoms. All the noble gases except helium have

Krypton Snapshot

Chemical Symbol:	Kr
Classification:	Nonmetal; noble gas
Properties:	Nonreactive, colorless, odorless, tasteless
Discovered By:	Sir William Ramsay in 1898
Atomic Number:	36
Atomic Weight:	83.80 atomic mass units (amu), sometimes rounded to 84
Protons:	36
Electrons:	36
Neutrons:	48, 50, 46, 47, 44, or 42 (in order of decreasing abundance)
State at 68° Fahrenheit (20° Celsius):	Gas
Melting Point:	–251°F (–157°C)
Boiling Point:	–244°F (–153°C)
Commonly Found:	In air, oceans, Earth's crust, and meteorites

eight electrons in their outermost shells. An atom with eight electrons in its outermost shell is unlikely to give up or accept electrons, making this the most stable arrangement for an atom. Helium is stable because its only shell is full with two electrons.

When these gases were first discovered, it was thought that they were rare. Thus, they were once called the rare gases. Then scientists discovered that some of them are not rare at all. Because they are generally nonreactive, they have been called inert gases. However, the term "inert" is also applied to gases outside this group—such as nitrogen and carbon dioxide (CO_2)—to indicate they are not flammable. Furthermore, as you will read later, compounds of krypton have been made.

The ability to form compounds means an element is not totally nonreactive. So, "inert gases" is not a very good name for the group either. The term "noble" has been used for centuries to describe a substance that does not react with oxygen. Since the gases in this group do not react with oxygen, "noble" is a good name for them.

Chapter Three
Getting and Using Krypton

Krypton is found in tiny amounts in oceans and in minerals in Earth's crust. It is also found in meteorites. In addition, the fission of uranium (U) and plutonium (Pu) in nuclear power plants produces both stable krypton isotopes and unstable radioactive isotopes. However, the most abundant source of krypton is the air around us. Krypton makes up about one-millionth of Earth's atmosphere.

Obtaining Krypton

We obtain krypton from the atmosphere through a process involving many steps. First, air is passed through filters to remove dust. Then, water and carbon dioxide are removed. The air that remains is compressed under high pressure, then cooled to a temperature of about −321°F (−196°C). At this temperature, most of the gases in air are in a liquid state. Any that are not liquid are removed. Then the liquid air is allowed to warm slowly. Different elements change from liquid to gas at different temperatures. As each element turns from liquid to gas, it is removed. This process is called fractional distillation.

Krypton and xenon remain liquid after other elements have become gases. Further fractional distillation separates krypton from xenon. The krypton is then purified by passing it over hot titanium (Ti) metal. The purified krypton gas is packed for shipment either in glass bulbs at normal pressure or in steel containers at high pressure.

Because the atmosphere contains so little krypton and isolating it takes so much work, krypton is expensive. It costs about two hundred times as much as argon, a noble gas that is much more plentiful in the atmosphere.

Krypton can also be obtained from the fission of uranium and plutonium in nuclear power plants. This source may become more important in the future if the number of nuclear power plants increases and if the plants continue to use fission rather than fusion to produce electricity.

Since krypton is expensive compared to other noble gases such as argon, its uses have been limited so far. In spite of this, scientists have found numerous applications for krypton.

These nuclear reactors in Byron, Illinois, are among more than one hundred in the United States. All the reactors together produce about 20 percent of the electricity used in America.

Krypton plays a role in airline safety. It is used in airport runway lights. Imagine landing an airplane at night without these lights!

Krypton in Business and Industry

Krypton has several applications in business and industry. It is used in some types of photographic flashes for high-speed photography. It is employed in high-powered electric arc lights and in the flashing lights along airport runways. Krypton is also used in some counters that measure X-rays. One important use of these counters is to detect X-rays that reveal the amount of sulfur—a pollutant—in coal.

One of krypton's radioactive isotopes, krypton-85, has several applications. It exposes flaws too small to be seen in metal surfaces. Krypton-85 collects in the flaws, and its radiation can be detected. In a similar way, krypton-85 is used to find almost invisible leaks in sealed containers. The radiation from escaping krypton-85 atoms is detected, revealing the leaks.

Krypton in Medicine

Several krypton isotopes have medical applications. Krypton-85 is used to identify abnormal heart openings. Because krypton-85 is radioactive, it gives off radiation, which can be detected by special machines.

Two additional radioactive isotopes of krypton—krypton-81 and krypton-79—are used to study lung function. Patients breathe in air

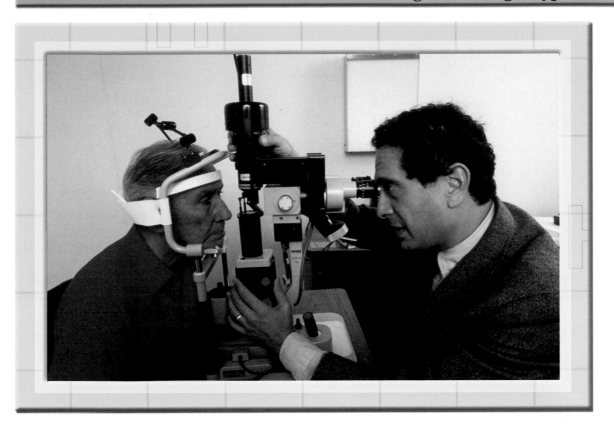

This doctor is preparing to use an argon-krypton laser to perform eye surgery on a patient. A brace strapped to the patient's head prevents him from moving during the surgery.

containing small amounts of one or both isotopes. A special camera takes pictures of the lungs by scanning the radiation from the isotopes. These pictures allow doctors to determine how well the lungs are working.

Krypton lasers help treat several eye problems. They can be used to repair a torn retina. They can also be used to stop bleeding and to slow or stop the growth of abnormal blood vessels inside the eye. In addition, krypton lasers are employed in facial plastic surgery.

Krypton can serve as an anesthetic under hyperbaric conditions. The patient breathes the gas at pressures greater than normal atmospheric pressure, forcing more krypton into the blood than would usually be the case. This reduces the oxygen in the blood and puts the patient to sleep.

The Krypton Meter

In 1960, the International Bureau of Weights and Measures defined the length of a meter as 1,650,763.73 times the wavelength of the orange-red light given off by krypton-86. Now that's a nice, round number! In 1983, the bureau replaced this definition of a meter with a new one. A meter is now defined as the distance light travels in a vacuum during 1/299,792,458 of a second.

Kryptonates

The word "kryptonate" probably has you thinking of Superman and kryptonite again. However, kryptonates are completely different from the fictional mineral that weakened Superman. Kryptonates are solid materials into which krypton-85 has been forced through the use of an electric field or very high temperatures and pressures. The solid material is usually a metal and is known as the host material.

Kryptonates can be used for a variety of purposes. They can be used to study surface temperatures of the host material, rates of chemical reactions, and surface wear. They can be used to detect even small amounts of dangerous gases such as hydrogen fluoride (HF) and hydrogen chloride (HCl). One kryptonate is extremely effective in detecting even very small amounts of dangerous, highly flammable hydrogen (H_2) vapor aboard space flight vehicles. Astronauts are sure to appreciate this use of kryptonates!

Krypton and Space Travel

Some scientists think krypton could be used as fuel for ion propulsion engines in spaceships. Ion propulsion engines work by passing electricity through gas in a chamber to create a plasma, which is a state of matter composed of ions. As ions stream out of the chamber, they move the spaceship forward. Ion propulsion engines are good for very long journeys in space because they do not require a lot of heavy fuel.

Ion propulsion engines are not powerful enough to overcome Earth's gravity and lift a spaceship into space. However, they would perform well for travel from the International Space Station, shown here, to the moon.

Other Uses for Krypton

Testing for radioactive krypton-85 produced at nuclear power plants can help protect plant workers. The testing can detect leaks in nuclear fuel containers that might endanger the plant workers. Measuring krypton-85 in the atmosphere can be used to monitor nuclear fuel reprocessing activities in other countries to make sure they are complying with the Nuclear Non-Proliferation Treaty, which limits the spread of nuclear weapons.

A Japanese researcher displays a million-year-old ice sample from Antarctica. It came from a depth of about 10,000 feet (3,048 meters). Scientists hope to learn a great deal about Earth's climate history from the sample.

Krypton-81 is a radioactive isotope produced high in the atmosphere by cosmic rays. It sinks from these high upper levels and is deposited everywhere on Earth. Scientists have used the isotope's rate of decay to determine the age of water contained in deep underground chambers. Knowing the water's age helps scientists understand what has happened to our planet in the past. Krypton-81 can also be used to date polar ice in studies of Earth's climate history.

Chapter Four
Krypton Compounds

Compounds are all around us. They are formed when two or more elements bond together. Compounds involve two main types of bonds: covalent and ionic. In covalent compounds, atoms share electrons. In ionic compounds, one atom donates one or more electrons to other atoms.

Right now you may be saying to yourself, "Wait a minute. Why are we discussing compounds? I thought the noble gases were nonreactive." Generally, that's true. Noble gas compounds do not exist in nature. However, they may form under the right conditions in a laboratory.

It took scientists a long time to discover this property of noble gases. Although noble gases were discovered in the 1890s, it was not until 1962 that the first noble gas compound was created. Compounds of xenon and radon were produced that year. The first krypton compound was formed in 1963. Before the discovery of noble gas compounds, however, noble gas clathrates were known.

Krypton Clathrates

Although clathrates involve the combination of two elements, they are not true compounds. No sharing or exchange of electrons occurs with clathrates. Instead, a molecule of one element, called the host molecule, forms a

Johannes Diderik van der Waals *(right)* won the 1910 Nobel Prize in Physics. He is shown here around 1919 in the lab of fellow Dutch physicist Heike Kamerlingh-Onnes *(left)*.

cagelike structure around a molecule of another element, called the guest molecule. Weak forces of attraction, called van der Waals forces, hold clathrates together. These forces were named for the Dutch physicist Johannes Diderik van der Waals, who first recognized their importance in the late 1800s. Clathrates are solids, even when the elements forming them are a liquid and a gas.

In 1957, scientists discovered that gases form clathrates with molecules of water or other liquids. Krypton forms clathrates with two liquids, hydroquinone [$C_6H_4(OH)_2$] and phenol (C_6H_5OH).

Although clathrates may sound like odd things, they do have practical uses. Suppose you want to obtain a purified sample of a substance in a solution. If the substance can form a clathrate with a gas, that property can be employed to obtain the purified sample, often more cheaply than other methods. The gas is added to the solution, where it forms a solid clathrate with the substance you want. The clathrate is separated from the remaining liquid, the gas is forced out of it, and the desired substance remains.

Krypton Difluoride

On March 23, 1962, British chemist Neil Bartlett created the first true noble gas compound. He combined xenon and gaseous platinum hexafluoride

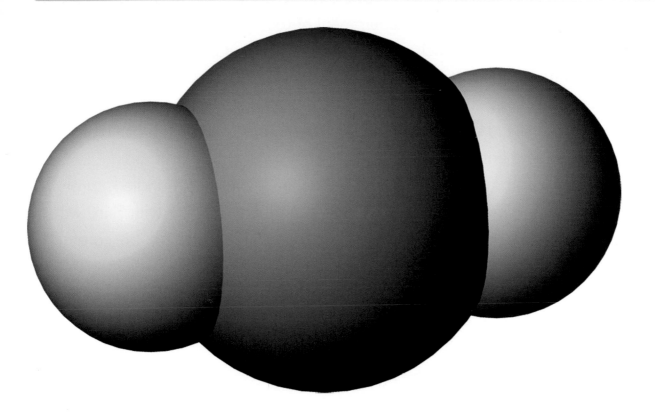

The atoms in krypton difluoride are arranged in a straight line—fluorine, krypton, fluorine. The krypton atom, with its atomic number of 36, is much larger than the fluorine atoms, whose atomic number is 9.

(PtF_6), which reacted instantly to produce a yellow-orange solid. That solid was the first proof the noble gases could indeed form compounds.

Bartlett's discovery quickly inspired attempts to form other noble gas compounds. In 1963, D. R. MacKenzie created the first krypton compound when he irradiated krypton and fluorine (F) in an electron beam at −238°F (−150°C) and produced krypton difluoride (KrF_2), a white crystalline solid. Krypton difluoride is still the easiest krypton compound to make and the most common one.

Krypton difluoride's most important use comes from the fact that it is a powerful fluoridating agent, a substance that transfers fluorine to another material. This property has enabled the compound to remove plutonium contamination from inside equipment at Los Alamos National Laboratory. Krypton difluoride reacts with the radioactive plutonium to form plutonium hexafluoride (PuF_6), which is a fairly stable gas and, therefore, easier to deal with than solid plutonium. Some scientists have suggested krypton difluoride could be used in a similar way to remove uranium contamination from equipment.

Although krypton difluoride is useful, it must be handled carefully. Because it is such a powerful oxidizer, it reacts violently with skin, water, and organic matter. Scientists using the compound must be careful to avoid contact with it.

Los Alamos National Laboratory, located in New Mexico, is part of the U.S. Department of Energy. It conducts research involving the application of science and technology to matters of national security.

Krypton-Uranium Compound

In 2002, scientists at Ohio State University and the University of Virginia announced the discovery that uranium could form compounds with three noble gases—krypton, argon, and xenon. As often happens with scientific research, the discovery came about by accident. The scientists were studying CUO, a compound of uranium and carbon monoxide (CO). They wanted to learn how radioactive metals react with small molecules such as those of carbon monoxide. For their research, they formed clathrates with noble gas molecules acting as host molecules for CUO. To their surprise, they found the noble gases were forming compounds with uranium atoms in the CUO. The scientists were so astonished, they didn't believe the results at first. However, continued testing confirmed the discovery. The scientists say this may open the door to further research. The same technique might be used to form compounds of noble gases and other metals. Perhaps more important, the scientists say the results provide a new window on how chemical bonds can form. New understanding of how chemical bonds form can expand our knowledge of chemistry as a whole.

Compounds of Noble Gases and Hydrocarbons

In 2005, scientists in Finland and Russia announced they had produced compounds of noble gases and hydrocarbons. Many scientists see this as a significant development. Hydrocarbons are the simplest organic molecules, incorporating only hydrogen and carbon (C). The enormous energy contained in hydrocarbons makes them very important to modern life. Hydrocarbons are found in crude oil and in fuels made from it, such as gasoline, diesel fuel, propane, and butane. Hydrocarbons also pose many risks. They help create ground-level ozone, which is a serious air pollutant that can cause breathing problems and lung damage. When hydrocarbons

Creating Fusion?

Scientists have long tried to produce energy by fusion. Fusion is what powers the sun. It is the energy released when two hydrogen isotopes are fused, or combined. Fusion has many advantages over fission. Unlike fission, its products are not radioactive. Its reactants, hydrogen isotopes, are abundant on Earth and could provide energy for millions of years. Some scientists are experimenting with using special lasers—including krypton fluoride (KrF) lasers—to produce fusion.

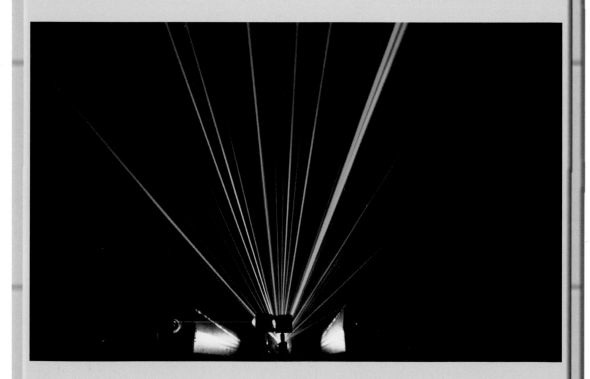

In a krypton fluoride laser, electric current passes through a gas containing krypton and fluorine, producing a concentrated beam of light. Multiple parallel krypton fluoride lasers are used in efforts to create fusion.

are burned as fuel, they generate carbon dioxide, a widely recognized greenhouse gas that contributes to global warming. Hydrocarbons can also cause cancer. Some scientists believe further research into hydrocarbon and noble gas compounds may yield important new life-improving products. For example, new anesthetic compounds might result. In addition, the hydrocarbon and noble gas compounds may lead to new fuels that are more energy efficient and create less pollution.

We have already looked at some of the uses of krypton and krypton difluoride. However, those uses are not part of the everyday lives of most people. So, by this time you may be thinking that krypton does not have much to do with your daily life. It may surprise you to learn that krypton is used in some objects that are so common, we rarely think about them.

Krypton and Lightbulbs

Lightbulbs are so familiar, we take them for granted. There are different types of lightbulbs, but most people use incandescent bulbs at home. This type of bulb has a thin tungsten (W) filament inside a glass globe. When electricity runs through the filament, it produces heat and light. Because the heat could cause a fire if there were oxygen inside the bulb, air is sucked out before the bulbs are sealed. However, this creates another problem. The heat causes tungsten atoms to break free from the filament. Without any gas in the bulb to slow or stop the tungsten atoms, the filament disintegrates, giving the bulb a short life. Since noble gases do not pose a fire risk, they are put inside the bulb to slow the filament's disintegration. Some tungsten atoms bounce off the noble gas atoms and return to the filament, making the bulb last longer. Low-cost argon is used in most incandescent bulbs.

More expensive krypton, however, is used in high-priced bulbs because its more massive atoms bounce more tungsten atoms back to the filament. This makes the filament last longer and allows it to be heated more, so it is brighter.

Incandescent bulbs waste a lot of energy producing heat instead of light. Nonetheless, people like them because their light has a lot of red in it and seems more like sunlight than the bluish light from more efficient fluorescent bulbs. They also are simpler and less expensive to manufacture.

Fluorescent Lights and Neon Signs

Fluorescent bulbs work differently than incandescent bulbs. A fluorescent bulb contains a noble gas and a small amount of mercury (Hg). Phosphors coat the inside of the bulb. When electricity flows through the bulb, it ionizes the gas and creates an electric arc that changes some of the mercury into gas. Some of the mercury gas atoms become excited. That is, some of their electrons get bumped to higher energy levels. When these electrons drop back to their normal energy level, they give off ultraviolet light. People cannot see ultraviolet light. However, it excites atoms of the phosphors on the

It is hard to imagine the commercial districts of the world's cities without the neon signs that light them up at night. This photograph shows the shopping district in downtown Seoul, South Korea.

Krypton for Headlights

Many cars have halogen headlights, which are incandescent bulbs with thinner-than-normal filaments that produce brighter light and let drivers see farther down the road. They get their name from the small amount of halogen gas—usually iodine (I) or bromine (Br)—they contain. Some halogen headlights also contain krypton. The krypton increases the pressure inside the bulb. This keeps the tungsten filament from losing as many atoms and helps the headlights last longer.

Headlights filled with the noble gas xenon are slowly replacing standard halogen headlights, such as the one shown here. The new, more expensive headlights produce a brighter, bluer light.

inside of the bulb, and they give off light people can see. The noble gas that is key to this whole process is usually argon. Sometimes, however, krypton is added to make the light brighter.

Noble gases are used in the familiar "neon" signs we see on many businesses. When electricity is sent through the signs' glass tubes, it makes the gases glow. Different gases glow with different colors. Neon, which gave its name to these lighted signs, produces an orange-red color. Krypton gives off a greenish-yellow color. Xenon produces a blue, and argon gives off a light bluish purple.

Krypton and Windows

As energy costs rise, people are trying to make homes and other buildings more energy efficient. Since windows play a big role in a building's heat loss or gain, improving their R-values and U-values will improve their insulating capabilities. The R-value is a measure of resistance to heat flow. The U-value is a measure of how much heat is conducted through something. The higher the R-value, the better. The lower the U-value, the better.

The air spaces between the panes of double- and triple-paned windows give those windows higher R-values and lower U-values than single-paned windows. A double-paned window insulates about twice as well

Rising energy costs have increased interest in insulating windows. It has been estimated that about 30 percent of energy-efficient windows sold in England and Germany use krypton in the spaces between panes.

as a single-paned window. A triple-paned window insulates about three times as well.

Using a noble gas instead of air in the spaces of double- and triple-paned windows improves their insulating abilities. Noble gas molecules are heavier than most molecules in the air, so they don't move as fast. Because they don't move as fast, they don't carry heat as well. Noble gases are usually used along with a special coating that reduces the amount of heat that travels through the window. As with lightbulbs, inexpensive argon is the most commonly used noble gas. Nevertheless, more costly krypton is used in some high-priced triple-paned windows. It is a more effective insulator because its molecules are heavier and slower than argon molecules. A triple-paned window with the special coating and krypton between the panes insulates about four times as well as an ordinary double-paned window and more than twice as well as an ordinary triple-paned window.

Krypton and Insulation

The development of energy-efficient windows led to the creation of a new insulating material known as gas-filled panels (GFPs). GFPs are sealed plastic pouches with a baffle—a honeycomb of small cells—inside. Filling the cells with a gas such as argon, krypton, or xenon provides the insulating power. Just as with windows, krypton gives higher R-values and lower U-values than argon. Xenon, with its larger, heavier atoms, offers even better insulation, but it is more expensive. GFPs work better than many traditional insulating materials and require less space. It is also easy to form them into any shape required.

GFPs have been used as packing material for cargo that needs to be kept at a low temperature, such as seafood, meat, fruit, and some medicines. They are less bulky and more efficient than polystyrene, which has often been used for such cargo. GFPs also do not raise the environmental and health concerns many people have about polystyrene, which

This photograph shows a GFP baffle. In the future, GFPs might be used to insulate cars. Their light weight and superior insulating abilities could lower the weight of cars and improve gas mileage.

critics say causes pollution and contains toxic chemicals that threaten human health.

GFPs can replace traditional insulators in ovens and refrigerators. Using GFPs in appliances can lead to energy savings. For example, a refrigerator insulated with GFPs could use up to 25 percent less energy than a foam-insulated refrigerator.

Right now, GFPs are too expensive to replace standard insulation in buildings. However, that may change as production costs come down. Some scientists also believe GFPs might be used in the future to insulate cars, leading to lighter, more energy-efficient vehicles.

We are really just beginning to know what krypton can do. As scientists find ways to obtain krypton more inexpensively, it may become much more widely used. Krypton may help us achieve a more energy-efficient future. We may find ways to use it that we have not even dreamed of yet.

The Periodic Table of Elements

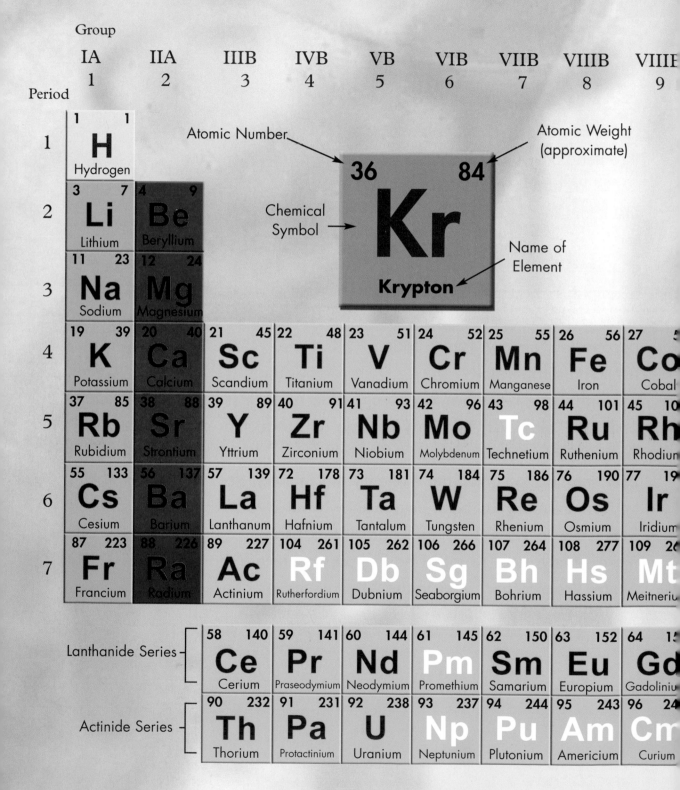

Group

| IA | IIA | IIIB | IVB | VB | VIB | VIIB | VIIIB | VIIIB |
| 1 | 2 | 3 | 4 | 5 | 6 | 7 | 8 | 9 |

Period

Atomic Number

Atomic Weight (approximate)

Chemical Symbol

36   84
Kr
Krypton

Name of Element

Period 1
1 1 **H** Hydrogen

Period 2
3 7 **Li** Lithium
4 9 **Be** Beryllium

Period 3
11 23 **Na** Sodium
12 24 **Mg** Magnesium

Period 4
19 39 **K** Potassium
20 40 **Ca** Calcium
21 45 **Sc** Scandium
22 48 **Ti** Titanium
23 51 **V** Vanadium
24 52 **Cr** Chromium
25 55 **Mn** Manganese
26 56 **Fe** Iron
27 **Co** Cobal

Period 5
37 85 **Rb** Rubidium
38 88 **Sr** Strontium
39 89 **Y** Yttrium
40 91 **Zr** Zirconium
41 93 **Nb** Niobium
42 96 **Mo** Molybdenum
43 98 **Tc** Technetium
44 101 **Ru** Ruthenium
45 10 **Rh** Rhodium

Period 6
55 133 **Cs** Cesium
56 137 **Ba** Barium
57 139 **La** Lanthanum
72 178 **Hf** Hafnium
73 181 **Ta** Tantalum
74 184 **W** Tungsten
75 186 **Re** Rhenium
76 190 **Os** Osmium
77 19 **Ir** Iridium

Period 7
87 223 **Fr** Francium
88 226 **Ra** Radium
89 227 **Ac** Actinium
104 261 **Rf** Rutherfordium
105 262 **Db** Dubnium
106 266 **Sg** Seaborgium
107 264 **Bh** Bohrium
108 277 **Hs** Hassium
109 26 **Mt** Meitneriu

Lanthanide Series
58 140 **Ce** Cerium
59 141 **Pr** Praseodymium
60 144 **Nd** Neodymium
61 145 **Pm** Promethium
62 150 **Sm** Samarium
63 152 **Eu** Europium
64 15 **Gd** Gadoliniu

Actinide Series
90 232 **Th** Thorium
91 231 **Pa** Protactinium
92 238 **U** Uranium
93 237 **Np** Neptunium
94 244 **Pu** Plutonium
95 243 **Am** Americium
96 24 **Cm** Curium

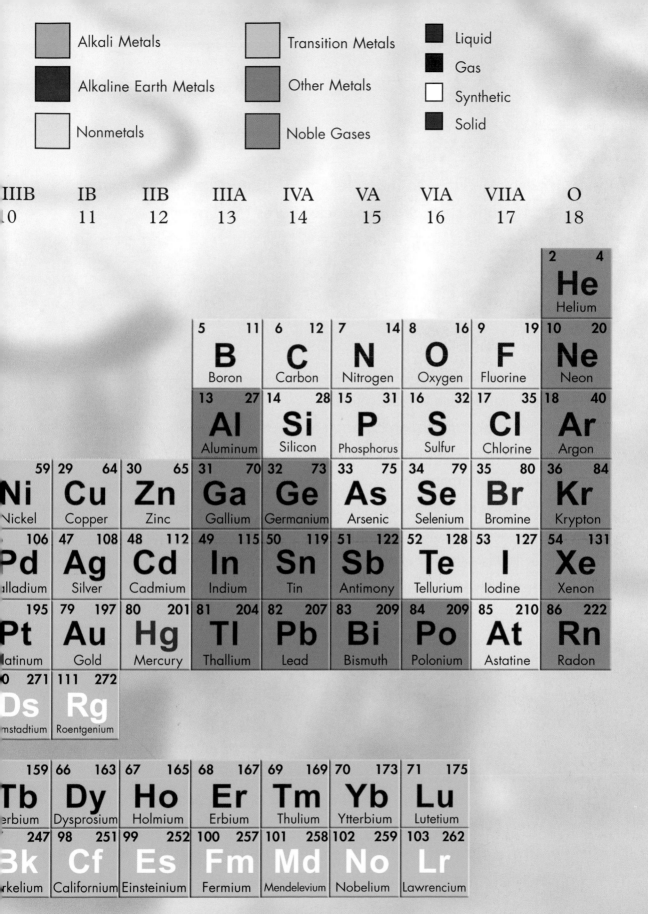

Glossary

anesthetic Medicine to make feeling or pain go away, or to make someone go to sleep while a doctor cares for him or her.

atom The smallest part of an element having the chemical properties of that element.

fission The breaking of a large nucleus into two smaller nuclei.

fusion The merging of two small nuclei into one larger nucleus.

inert Not chemically active.

insulate To prevent the transfer of heat.

irradiate To expose to radiation.

isolate To separate one substance from another.

isotopes Atoms containing the same number of protons but different numbers of neutrons.

laser A device that produces light of a single energy by forcing electrons in atoms or molecules to jump from a specific high-energy state to a specific lower-energy one.

molecule A group of atoms that are chemically bonded together.

organic Having to do with living organisms.

phosphor A substance that gives off light when excited by radiation.

polystyrene A type of plastic that can be made into a foam that is a good insulator and can be molded into different shapes. Styrofoam is a brand of polystyrene foam.

radioactive Having the property of giving off subatomic particles as the result of the disintegration of atomic nuclei.

radioactive decay The breaking apart of the nucleus of an atom, forming a different element.

retina The innermost layer of the eye that is very sensitive to light. It is connected to the brain by the optic nerve.

For More Information

American Chemical Society
1155 Sixteenth Street NW
Washington, DC 20036
(800) 227-5558
Web site: http://www.chemistry.org
The American Chemical Society is the national organization for professional chemists. It also provides information about all aspects of chemistry for students and educators.

Efficient Windows Collaborative
Nils Petermann
Alliance to Save Energy
1850 M Street NW, Suite 600
Washington, DC 20036
(202) 530-2254
Web site: http://www.efficientwindows.org
The collaborative is a gateway to information about energy-efficient windows, including multipaned windows with noble gas fills.

Environmental Energy Technologies Division
E. O. Lawrence Berkeley National Laboratory
1 Cyclotron Road
Berkeley, CA 94720
(510) 486-6784
Web site: http://eetd.lbl.gov
The Environmental Energy Technologies Division is a research facility dedicated to research on energy technologies and efficiency, and the

environment. One area of research is energy efficiency of buildings, including lighting and windows.

International Union of Pure and Applied Chemistry
IUPAC Secretariat
P.O. Box 13757
Research Triangle Park, NC 27709
(919) 485-8700
Web site: http://www.iupac.org/dhtml_home.html
The IUPAC is an international, nongovernmental organization dedicated to advancing chemistry worldwide and contributing to the application of chemistry in the service of humanity.

Jefferson Lab
Office of Science Education
628 Hofstadter Road, Suite 6
Newport News, VA 23606
(757) 269-7567
Web site: http://education.jlab.org
Jefferson Lab is a world-class research facility with a commitment to science education. It provides considerable material concerning many aspects of chemistry.

Web Sites

Due to the changing nature of Internet links, Rosen Publishing has developed an online list of Web sites related to the subject of this book. This site is updated regularly. Please use this link to access the list:

http://www.rosenlinks.com/uept/kryp

For Further Reading

Curran, Greg. *Homework Helpers: Chemistry*. Franklin Lakes, NJ: Career Press, 2004.

Ganeri, Anita. *Neon and the Noble Gases* (The Periodic Table). Chicago, IL: Heinemann Library, 2004.

Keller, Rebecca W. *Chemistry: Level 1* (Real Science-4-Kids). Albuquerque, NM: Gravitas Publications, Inc., 2005.

Miller, Ron. *The Elements: What You Really Want to Know*. Minneapolis, MN: 21st Century, 2006.

Newmark, Ann, and Laura Buller. *Chemistry* (Eyewitness Books). New York, NY: DK Publishing, 2005.

Newton, David E., and Lawrence W. Baker. *Chemical Elements: From Carbon to Krypton*. Farmington Hills, MI: UXL, 1999.

Oxlade, Chris. *Elements and Compounds* (Chemicals in Action). Chicago, IL: Heinemann Library, 2002.

Stwertka, Albert. *A Guide to the Elements*. 2nd ed. New York, NY: Oxford University Press, 2002.

Tocci, Salvatore. *Hydrogen and the Noble Gases* (A True Book). Danbury, CT: Children's Press, 2004.

Wertheim, Jane. *The Usborne Illustrated Dictionary of Chemistry*. 2nd ed. London, England: Usborne Books, 2000.

Bibliography

Chleck, D., R. Maehl, and O. Cucchiara. "Development of Krypton 85 as a Universal Tracer." Energy Citations Database, 1962. Retrieved August 22, 2007 (http://www.osti.gov/energycitations/product. biblio.jsp?osti_id=4172694).

Freudenrich, Craig C. "How Atoms Work." HowStuffWorks. Retrieved July 27, 2007 (http://science.howstuffworks.com/atom.htm).

Gartner, Ted. "Technology Transfer: Gas Filled Panels." Lawrence Berkeley National Laboratory Technology Transfer Department. *EETD* Newsletter, Vol. 6, No. 1, 2005. Retrieved August 29, 2007 (http://eetdnews.lbl.gov/nl20/tt.htm).

Gorder, Pam Frost. "Chemists Make First-Ever Compounds of Noble Gases and Uranium." Ohio State Research News. 2002. Retrieved July 23, 2007 (http://researchnews.osu.edu/archive/noblegas.htm).

Hebrew University of Jerusalem. "New Noble Gas Chemical Compounds Created." ScienceDaily, 2005. Retrieved July 23, 2007 (http://www. sciencedaily.com/releases/2005/03/050323115810.htm).

Kauffman, George B. "Sir William Ramsay: Noble Gas Pioneer—On the 100th Anniversary of His Nobel Prize." *Chemical Educator*, Vol. 9, No. 6, 2004. Retrieved July 25, 2007 (http://chemeducator.org/ sbibs/s0009006s/spapers/960378gk.htm).

Koppe, Karsten. "Synthesis, Reactivity, Structural, and Computational Studies of [C6F5Xe]+ and [C6F5XeF2]+ Salts." Ph.D. diss., Universität Duisburg-Essen, 2005.

Los Alamos National Labs Chemistry Division. "Krypton." Retrieved July 23, 2007 (http://periodic.lanl.gov/elements/36.html).

Thermal Industries. "Windows." Retrieved August 30, 2007 (http:// www.ahc-inc.net/Windows/Thermalindustries/ThermalInd.htm).

About the Author

Janey Levy is a writer and editor who has written numerous books for young people. Her interest in chemistry comes from her curiosity about the role the body's own chemistry plays in an individual's physical, emotional, and mental well-being. Levy lives in Colden, New York.

Photo Credits

Designer: Tahara Anderson; **Editor:** Kathy Kuhtz Campbell
Photo Researcher: Amy Feinberg